In Search of Gems

Finding Treasures in Wild Places

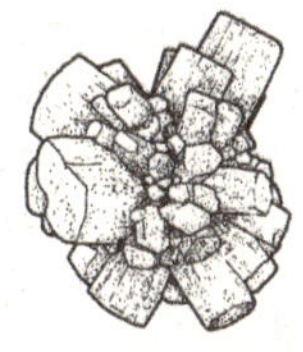

IN THE MOMENT

Kenneth Steven

Saraband

Published by Saraband
3 Clairmont Gardens
Glasgow, G3 7LW
www.saraband.net

ISBN: 9781916812628

Printed and bound in Great Britain by Clays Ltd, Elcograf S.p.A.

1 2 3 4 5 6 7 8 9

Contents

Introduction 1
Serpentine 7
Mica 18
Garnet 27
Amber 35
Pearls 45
Aquamarine 55
Smoky Quartz 64
Gold 73
Agate 81
Diamond 92

Appendix: On gem collecting 99
About the author 104

Praise for Kenneth Steven's previous writing:

"Offers a warmly inclusive and decidedly secular series of essays about what it means to make a pilgrimage in the early years of the 21st century." —Roger Cox, *The Scotsman*

"Beautiful and evocative." —*Kaggsy's Bookish Ramblings*

"A lovely little gem, contemplative and peaceful." —*Belle Reads Books*

"Kenneth has a rare gift of being able to transmute the mundane into the mesmerizing, in a kind of poetic alchemy." —*Countryman*

"Enjoying Kenneth Steven's work intensely – style and subject." —Ted Hughes

"Kenneth Steven has a ready sensitivity to the beauty of small moments." —*The Herald*

"There is a grave beauty in these lines, revealing a poetic voice of great sensitivity." —Alexander McCall Smith

"Steven has a talent for capturing the startling, original image … he is a fine, fine poet." —*New Shetlander*

"This complex picture … will leave readers unsettled." —*Publisher's Weekly*, starred review

"Impressive … offers a short, sharp reminder of how delicately balanced many modern societies actually are."—*Booklist*

This book is dedicated to the memory of Robin Field, composer, who with his wife Jean patiently taught me to find agates.

“But still the Vine her ancient ruby yields,
And still a Garden by the Water blows.”

The Rubáiyát of Omar Khayyám

“These gems have life in them: their colours speak, say what words fail of.”

George Eliot

Introduction

A great deal of this book is made up of the stories of how I first became excited by Scottish stones, especially gemstones. There are several roads in to knowing why I first developed this real fascination, and the truth is that I don't know for certain now which was the most significant. The fact of the matter is, though, that it has remained a very real interest and indeed a love into adulthood, but it did not result in me deciding to study geology. Nor even did that even cross my mind. I freely admit that my passion ultimately is for gems rather than minerals.

The very word 'mineral' conjures up some grey slab of stone that, however interesting it might be to a true geologist or scientist, strikes me as fairly dull. From a young age, I was fortunate enough to encounter some amazing gemstones, and these encounters made me the magpie I am today. Stones for me have to possess some kind of classical beauty: breathtaking colour or fabulous facets. Knowing where they come from matters too, but that's of more recent importance to me.

I write in these chapters about what I stumbled over in my childhood and adolescence, even into early adulthood. I suppose it was about more than simply good fortune; if you're fascinated by *any* subject, you want to dig more deeply into it, and in the old days that had to be through books, or through the experience and knowledge of other people. And because of that early fascination with gems, I was given stories and folk shared their accounts with me.

I still don't believe I truly realised what a treasure trove of gemstones Scotland has before I was lent a little book, *Scottish Gem Stones* by W.J. McCallien, first published in 1937. I have to stress the word *lent* here because it was not a book that I was given.

I am rather haunted by the wise words of Shakespeare: *Neither a borrower nor a lender be.* The person who lent me the book, because she came to know of my absolute addiction to gems, had herself been a geology student at the University of Aberdeen. To some extent, it's true that her days for hunting east coast beaches for pieces of agate or amber were done, but I'm conscious of sounding as though I'm making excuses for myself. I do remember when she first gave me this most insignificant little book with its decidedly dull cover. Although it was compiled by a former lecturer in geology from the University of Glasgow, the work it contained was very much detailing the locations mineralogist Matthew Heddle (1828–97) had visited so meticulously. During his lifetime, he compiled the greater part of his *magnum opus, The Mineralogy of Scotland*. In doing so, he must have tramped over every locality in every possible type of terrain; little surprise that because of his thoroughness, he uncovered a great many new localities too. But Heddle didn't actually manage to complete the book before his death. It was then edited by another eminent Scottish mineralogist, John Goodchild, with the help of Alexander Thoms. Like Heddle, Thoms was a collector of Scottish minerals, and doubtless he came to know the great man well; after all, Heddle was his father-in-law. I return to something of the story of Thoms in the last chapter.

As I mention, I still remember being given this little book and scouring its pages with utter joy and a pounding heart. I found location after location that I wanted to visit right away, as soon as I could leave my home county of Perthshire. I wanted to visit Sutherland glens in search of gold, and scout in particular for aquamarine and topaz. Some of these locations I have visited, and those experiences are related here; others remain very real objectives.

The difficulty is, as I mention in one or other of the chapters, that often it's not sufficient simply to know a location (whether an island or a glen or a river). You really need to have the precise grid reference for whatever the treasure might be. But as I mention too, the flip side of that is that it creates the treasure hunt. There's a need to go and search for yourself, and often enough in a beautiful corner of Scotland: hardly a hardship. The journey can be as much a part of the joy as the actual finding.

Some of the chapters in this book grew into essays for what became a series on BBC Radio 3. I had written a number of earlier series of essays for radio, which were edited and produced by my good friend Mark Rickards, for long years a producer with BBC Radio. I had worked for several years too with the Publisher at Saraband, Sara Hunt. Saraband brought out the book I worked on with my wife, Kristina, on the cultural history of the Norwegian Sami, *Beneath the Ice*, and a more recent volume of Scottish essays entitled *Atoms of Delight*. It was Sara who, on discovering my passion for gemstones, requested that I write this book. I realised that I had some memorable stories describing the roads I'd gone down because of that first passion for all that glistered. I'm grateful to Sara for offering the idea and for her and Heather Merrick's work as editors, as I am to Kristina for having read the chapters and suggested amendments.

I've still got that most insignificant-looking little brown-covered book in my library. I'm determined yet to discover all manner of locations for Scottish gems. It's the kind of book you bring out by the fireside in December, when the winds are whelming about the place, to dream about the journeys there might be the following June.

Serpentine

The great Scottish mineralogist Matthew Heddle, who lived through the middle part of the nineteenth century, gave me the name of many a location for my hunting of semi-precious stones in my adolescence. I was growing up in Highland Perthshire, the only truly land-locked part of mainland Scotland. It's an area blessed with an abundance of lochs, many a Munro (that's a mountain over 3,000 feet), great rivers and more, but it doesn't have the sea. And I grew up exploring all this wildscape, being taken out to walk wild glens, to visit places off the beaten track. And while my parents had their eyes fixed on the skies in their eternal quest for birdlife, mine were focused on the ground beneath my feet in the hope of spying gems, whether smoky quartz or flint or agate.

But there was one gemstone I did know long before I'd stumbled with gratitude on Heddle's listings of locations for Scottish stones, and that was serpentine. It's a strange-named gem in my mind because its markings are supposed to be reminiscent of the skin of a snake. Well, they're not like any snake I've encountered to date. The first thing that can be said with certainty about the outside of a pebble of serpentine is that it possesses a waxy feel. It's a very soft stone, making it easy for jewellers to work with. But the stones I've encountered over long years (for I've been a hunter for serpentine since I was five years old) are commonly white (that's the quartz part) with yellow or green blotches. It's almost as though a child had taken a paintbrush to them to make them spotty.

The ones I've been hunting for ever since I learned the tricks of the trade are pebbles of dark green (I like to call myself teasingly a green supremacist). It's rare indeed that a pebble will be nothing more than green (either the shade of lettuce or as dark as jade) because generally it will have a white spot here or there. Sometimes too a green fragment or pebble of serpentine will be fully translucent. These stones are going to be prized most by the jewellery makers for rings and brooches and whatever else. Serpentine can be so soft, though, that I've known a long slice of stone actually to snap between my fingers.

I think I must have found serpentine in every corner of the island of Iona; by that I mean on every beach. I should explain that Iona is in the Inner Hebrides off the Scottish west coast, lying just west of the long foot of the neighbouring island of Mull that kicks out into the sea, for all the world like some footballer's long leg. Iona itself is of no great size, being about three miles long and just one and a half wide. Its fame is so great because it was here that St Columba landed from Ireland back in 563 AD with his twelve followers. It was on the east side of Iona that Columba and those followers built their monastery, and from here that they went out with the flame of Christianity to convert the mainland. Three hundred years later, the Book of Kells, that great treasure of the Celts, was begun here on Iona, but because of Viking raids was whisked back to Ireland for safety and completed there.

But this tiny island became the beating heart of the Celtic Christian world in those early centuries. The monastery grew into a kind of early university where ideas of every kind (by no means only religious ones) were argued over hammer and tongs. It was a vibrant place of creative thought; people came here because they'd heard of Columba, and after his life was over they wanted to be buried in Iona's holy soil. No one knows just how many Scottish and Norwegian kings may have been laid to rest here, but it's a considerable number.

The holy bit of Iona, if you like, is on the island's upper east side. And it's hardly any accident that should be the fertile part; a monastery needed both fields and a garden. The east side too is more sheltered from the prevailing winds, and they can be a force to be reckoned with. The inland part of Iona is divided between lush green glens and meadows, and gnarled hummocks and hills of grey granite.

I always wonder if Columba and his monks knew about serpentine, because oddly enough, the very best pebbles are to be found on the beach where their *curragh* landed from Ireland. Perhaps they were too busy, for that romantic notion that somehow this was a haven of retreat is arrant nonsense. These were warrior monks, and Columba is reputed to have been fierce. They spent their days building and carrying and digging; nights praying and singing and writing, for there was a scriptorium here. But had they time for serpentine? I like to think they were aware of it, for those early Celts had eyes for beauty and the celebration of beauty, both that which they created and that which they saw about them.

The expedition to St Columba's Bay (for there was only one each time we visited the island) was a high point for me, and from the age of five it was because of the serpentine hunt that was to happen.

Of course, the packing and the preparation by my parents in readiness for the day all seemed like wasting time. I wanted nothing but to get there and begin the hunt. To explain, we'd be walking from the straggled line of cottages that composed the village on that sheltered east coast to the south end of Iona. That whole southern half of the island is characterised by gnarled granite and wildscape. It feels a place that hasn't changed since the days of Columba and long enough before. Each side of this southern half is studded with tiny coves, some so difficult to reach you take your life in your hands scrabbling over boulders and negotiating rough sheep tracks.

Eventually, the preparations would be done and we were ready to leave for the bay. My memory now is of my father and mother carrying however many bags and my father always with binoculars round his neck. I scampered ahead on the track, too impatient to wait for them, then scampered back to see if they couldn't walk more quickly, for all the world like some impatient sheepdog. Now at last we'd come to the island's west side and you could hear the roaring of the sea as it combed the beaches and drove against the headlands. We turned south and were into what I like to call the great loneliness of Iona, a kingdom of ravens and otters and some of the oldest rocks in the world. I think whenever I pass that it might be a landscape from the Old Testament where once the prophets roamed.

Then, at last, we climbed a steep and rocky hillside that my parents nicknamed the Khyber Pass. From the top, you could turn and see all the broken fragments of numerous other islands in the Inner Hebrides to Iona's west. But I had no time for that because now, ahead of us, Columba's Bay was visible, just and no more. I always asked if I could run from here, down into a green glen that was lit with wild flowers in the summer. But I had little interest in them either; I was running so fast I felt I might fly until, finally, my heart hammering, I was down on the pebbled shore, my eyes searching already for flashes of green stone.

*

It was my mother who'd taught me to search for serpentine. It was she who had explained to me the difference between these waxy, much softer pebbles and stones made of marble. Those are duller; they don't polish the same, and once out of water they have a drabness about them. You can tell a piece of serpentine because as soon as it's absorbed the oils from your hand, it's polished, glimmering.

My mother was a sifter. I see her always when I visit all this time after her passing, sitting carefully working through the pebbles beside her in the hope of finding a green treasure. I was too impatient for that; I had to be a tide dancer, creeping the length of the beach closest to the waves to catch what might have been washed in. Always I'd get one drenching, having dared to grab at a green stone as one particularly mighty wave came crashing in.

I've said St Columba's Bay was best for serpentine, but those many other coves at the south end of Iona are worth visiting too. They can also hide pieces of green treasure, and often simply because they're less visited. But getting to them is not for the faint-hearted; they're isolated and lonely places. It's worth looking for the Marble Quarry on Iona's east coast, though it's maddeningly difficult to find (my cousin always maintains they move it every year). The marble is white with beautiful flecks of green serpentine through it. On no less than three occasions, there were attempts to create a commercial enterprise, but ultimately none was successful. It's not actually all that large a vein of marble. The Duke of Argyll evidently purchased and polished a piece back in 1693 (in those days, Iona belonged to his family).

The problem from the beginning was accessing this remote little cove, for to this day there's not a road of any kind that leads here. So the rock was taken away by ship and Iona marble went around the world. Today, those with a fascination for industrial archaeology will want to look at the very rusted remains of a large winch and cable and a cutting frame. A few cut blocks of marble remain. Those huge blocks had to be cut by hand, since dynamite would have destroyed the soft rock. The old marks of chisels and saws can be found on the great slabs that remain.

I think what I find most poignant of all are the remains of the quarrymen's cottage. Living and working at the Marble Quarry must have been a hard existence indeed; this place is no less remote today than it was back then. There's not even a source of fresh water. All of what remains is a rather eerie reminder of a project that didn't last. The quarry closed for the final time at the end of the First World War. But you can still locate a magnificent slab of the polished white granite with veins of the famous serpentine: go to Iona Abbey to see the altar standing there.

For myself, I'd choose serpentine over marble anytime. It's the uniqueness of the patterning on each stone, and the fact, too, that the range of the serpentine's shades is so diverse. I've alluded to the lettuce and jade green colours, but I haven't mentioned they can have the most beautiful blues and oranges and even reds too, not unlike the serpentine that's found at the Lizard in Cornwall.

Every time I leave St Columba's Bay at the end of another day of hunting, I try to remember to do something I learned from the Iona Community, the group that came here first back in the 1930s to rebuild Iona Abbey and to follow as they saw it in the steps of Columba on that first Celtic Christian path. They will invite you to throw a stone back into the sea, representing something from your life you want to leave behind. It's not just the bag that seems less heavy then but somehow the heart as a whole.

Serpentine

She knew to go on the storm days;
dawns after the wind had howled its worst.
She was up long before light
out into giddying breeze, the sea chalked
here and there on the distant rocks.
She'd only one frail hour
before work at the monastery began.
For that reason she ran, her feet knowing the path
in the half darkness of morning's gloom.
Beyond the monastery not a single soul
just the dun warmth of cows,
their large curves gathered at the field's end.
She ran all the way to the cove her father had shown her
when she was a child, cliffed on each side;
just a single horseshoe of shingle beneath.
Now the running was done
she went down hand after foot,
slowly and carefully, knowing the way.
Then those few steps to the beach and she bent
to know what the storm tide might have brought,
and somehow she still was a child, heart hammering hard.

Serpentine

Yet nothing at first: brown stones, then seaweed,
one pure white stone and some gravel. A wave
toppled in,
creaming the sand where her feet had just been.
She went back, saw the light was becoming
moment by moment more. She had minutes
before she must run the whole way back north.
Then there, at her feet, a green shining
the size of her pinkie nail. Polished already and
perfect.
She reached it before the sea returned.
That was the gift she had come for:
she hid it safe in the folds of her skirt,
then was up and away to her work.

Mica

My mother's family was Highland through and through. On both sides, they came from that belt of country stretching from Achnasheen in the west to Dingwall right on the other side, so from the wilds of Wester Ross over to the heart of Easter Ross. That belt's north of Inverness, and the line I've delineated follows the route of the West Highland Railway pretty much exactly. At one time, both sides of that family must have been Gaelic-speaking; now in the years long after the Clearances and the wars, the Gaelic was patchy and its presence hard to predict. It was strange, for example, that my grandfather who had grown up in the wilds of the country around Achnasheen should have known barely any Gaelic at all, while my grandmother from Strathconnon would have been a good speaker of the language and possibly a writer of it too (certainly, her mother was both). But that had to do with the attitude of local schoolteachers and headmasters as much as anything else; over the years, I've spoken to numerous Highlanders and islanders who will surprise me either with their accounts of how they were persecuted for having Gaelic, or else how they were encouraged with the language. The truth is that it's a complex map of different experiences.

The sad fact is that I missed out on meeting almost all these relations because my parents were older and consequently their parents and most of their relatives had died some years before I was born. I knew my great aunt Darla in Muir of Ord and she became a surrogate grandmother to me, and I knew one of her surviving brothers living still on a sheep farm with his wife. That was a glimpse into an older world if ever there was one. I remember yet the smell of the farmhouse and of the sleek short-haired collies that lay by the peat fire. What I remember more than anything is the sword from the Battle of Culloden that hung above the fireplace in the living room. My great uncle had found it in a peat bog somewhere on the land and it was so well-preserved by the peat you could read still the shining letters of the name of the English officer that once had borne it.

I did encounter any number of my mother's childhood friends. That was around Loch Ness, for she and her sister had grown up in Drumnadrochit.

Often, the friends we visited lived up steep roads rising from around the loch. It was a strange mixture of country; here we were ostensibly in the Highlands proper, yet it was by no means always *high land*. Around Dingwall, for example, where my maternal grandmother had grown up, the countryside was comparatively lush and fertile. In autumn, there were gold fields of barley and patches of woodland (through which a river famous for pearls flowed). Where my grandfather, my mother's father, had been raised far west at Ashnasheen, it could hardly have been wilder. There it feels like the Highlands proper, and it changes between the two across this belt of country, sometimes all at once and with no warning. Garve itself, on the road to Ullapool, and an important place to many of my mother's family, was suddenly into wildscape, but only a handful of miles away the land was farmed and flat. At any rate, what I remember most are the high farms and crofts. I would sit there content for a time, drinking tea that was dark as peat, but then I would grow restless and yearn to be set free to explore all that lay around that place. We would be there first each year in the Easter holidays and I think back now to the cold of those days.

The wind had a knife to it yet and there would be patches of snow in the hills around us. But the cold didn't matter one bit; I wanted to be out and exploring. There were few dangers; as I sit here writing now, I struggle to think of any at all. None of these hill roads were busy, but I knew well enough to be careful of cars anyway. There were certainly adders in these Highland corners, but with me skipping about and singing, they were more likely to have been frightened of me and I never did encounter one. No, there were few concerns for my safety and I could be out free range as long as I pleased, or until the cold drove me back inside. I want to give a sense of the kind of places these were.

Usually Loch Ness would be down below, at the bottom of steep, steep hillsides. But up there, perhaps a thousand feet and more above sea level, there'd be all manner of other lochans dotted about in the folds of the hills. I remember that Highland hill country being strangely pale compared to what I knew further south, and I'm not quite sure why. Almost a sandy shade, as though it was drier. A few trees and juniper bushes; bog myrtle and bog cotton. There was often a great deal of gorse too; I liked getting the sweet scent of it when I squeezed the leaves. That was true of the bog myrtle too, its aroma lovelier than the gorse. There might be a few very ancient pines, and up at that height they'd be sculpted by the wind. Often, there'd be a huddle of ancient pines by the lochs we visited. But generally the hills were bare; the ground around those crofts was stony and dry, and what I remember more than anything were the boulders. There always seemed to be such boulders up on this higher ground, strewn across it, almost as though once upon a time giants had fallen out and taken to hurling rocks at one another. These were the remnants of battle.

And they glittered, all of them, whether huge as a house or smaller than I was; they shone almost as if they contained small mirrors. And when outside, away from what I felt then was endless talk and tea, this is what excited me most. Mica. For that was what made up those tiny mirrors in the granite boulders, and some were very far from being tiny. This is where I wish I had greater knowledge of geology, but I've gone back to the Heddle book for assistance. The best I can do is offer as clear a picture as I can. To Heddle, mica was so common in Highland Scotland he didn't suggest particular sites for finding it; he did that instead for all the real treasure-hunters' gems – agate and tourmaline, carnelian and onyx, jasper and amethyst, all those aquamarines and topazes. But he must have encountered so much mica in his days of leaping over moors and hills there was no particular need to mention it.

It was certainly far more common here above Loch Ness than back in Perthshire. Again, that's down to geology, and again, I don't know why it should have been so prevalent here. I certainly remember catching odd glints of mica when I was up in the hill country above Aberfeldy, but it was glints rather than full pieces. Here in my mother's home country, it was shining at me from everywhere and it would have taken several lifetimes to find every bit I caught sight of in just the locations we visited. But just as common as the glints of it were the larger mirrors in those rocks the giants might have warred with. Those pieces could have been as much as half the size of my small hand back then; they were real chunks of the stuff, silvery and opaque. They had a kind of glassy sheen to the surface. What was wonderful to me was that often enough, they could be chipped and tugged from the granite rocks that held them – I suppose you might think of them as eyes in a large potato – so that one large boulder might keep me busy the whole time my parents were talking down at that fireside.

The mica itself was soft indeed, so you could have scratched the surface of one of those 'windows' with a penknife. If I managed to get out a whole chunk intact (and that was always the objective), I would break off the crusted bits of brittle stone around its edges. But there was more work to be done with the mica once it had been freed from its rock. It's one of the most cleavable of all minerals, so it can be split into any number of thin sheets. Indeed, the mica was so soft I could do the splitting with a single fingernail, though I seem to remember it left the said fingernail rather jagged. What I recall too, time and again, was holding up a thin sheet of mica to the light. I could see the hills and the sky through it, but almost like looking through a windscreen at the view beyond as the heaviest rainstorm drives against it. The world looked yellowish through them too; a smoky yellow.

I suspect many walks on good spring and autumn days in the Highlands must have been frustrating to my parents, for I would have been racing from boulder to boulder in my search for hoped-for patches of mica. Always, the dream would be to find one that was bigger than the last, and then of course I'd have to stop and try to chisel this one out while my parents waited impatiently. I can imagine them almost hoping for the wetter days when even my ardour for stones was dampened and I could be kept at heel as we followed the hill path to a loch.

The strange thing is that not a single piece of mica survives from those years. I do know what must have happened in the end: they simply broke with too much handling. I would have polished the surfaces of those chiselled-out mirrors of mica, and after a time, they snapped. I'd put a whole piece carefully deep in a pocket, one I'd managed to extricate from the granite. It would be the treasured piece of the moment and then sure as fate, it would break. Then another one would replace it after however long, and another, and so on.

What I think back on now is just how important those hunts to find mica were at that time; my child's mind was occupied completely by them. The cold of the day most likely there in early April, high in the hills; that meant not a thing. I had to find mica, and as big a piece as possible, and in my mind I'd be thinking too about those to whom I'd show the latest treasure. It always seemed important to share them with other eyes. There I would crouch on the hillside, digging away to get the biggest piece I could find with trembling hands as sometimes a great gust of sleet came and the glen's light was gone and all the hills swallowed. They might even be calling for me down at the croft house, telling me it was time to go now, but I couldn't tear myself away, not yet. I'd shout back instead and tell them I'd be there soon, once I was done, but that I couldn't come yet. Not quite yet.

Schist

I heard a story once of a boy in Iceland
searching the moorlands in the drizzling
spring,
hoping always for a raven's nest.

Finding in the end that straggled heap of sticks
and in among them, underneath, strange
stones
that flickered, winked with mica.

Brought back in the bird's own beak;
strange treasures for that grey abandoned place,
since every crow loves shining.

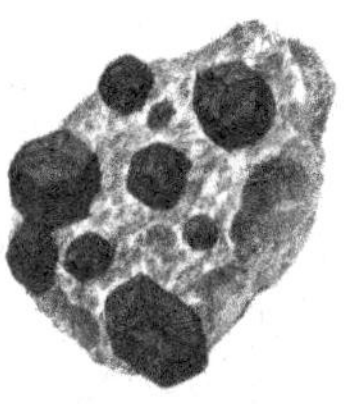

Garnet

Visiting the McTaggart family was always an adventure. They lived on a farm a thousand feet or so above Aberfeldy, at the heart of Highland Perthshire. I think there were nine children when I started visiting and eleven later on; the truth is it was such a large flock with babes being comforted in some corner of the house and one of the children suddenly rushing out for eggs that trying to count the whole lot at any one time was a challenge. I was an only child and came from a very quiet home. It was a revelation to me that such a large family could live together harmoniously, for certainly they did. I don't remember a single argument, though I do recall many a gale of laughter. If the inside world was busy, the one outside was little different. This was a croft rather than any real farm; it was subsistence farming. The children were extremely keen to see me trying to milk the goat; I still remember their shrieks of mirth at the fact I couldn't get a single drop from the beast, probably because it didn't know and trust my hands. There were many hens, a few sheep, a cow or two perhaps, one very vocal Siamese, and a snuffling dog called Georgia.

The sledging we did together was some of the best I've experienced in Scotland, for up at a thousand feet the snow was four feet deep, not the slippery marzipan mess that it was down in Aberfeldy. I remember once the sledge I was on taking off and the three of us landing in a whirling drift of snow. When we looked back, we saw we'd skimmed over the top of a barbed wire fence. In summer, there were expeditions to swim, for I remember well how very hot and muggy the house became even there so high above the valley. The swimming was for the older children only; the small ones remained at home, all the windows wide open as the parents tried to draw in some fresher and cooler air.

The father was the grandson of a highly respected Scottish painter. When I think of his work now, what comes to mind are depictions of children playing on beaches and his ability to catch the movement of the sea. Money truly seemed to mean nothing in the least to these McTaggart parents now with their seemingly ever-growing family. I know that once the older children took me up to the attic to show me one of the grandfather's original paintings and

I held it. I remember the framed painting covered with dust and pretty much picked up from the floor. It could have gone for a high price indeed to any gallery in Scotland.

But of all the many adventures we had, the most exciting was the day we had our expedition to Farragon. The hill almost sits in a triangle right above Aberfeldy, over a hillscape of moorland and small lochs, so to reach it we had to walk east from the farm, across that wildscape, and then climb the triangle. I think there would have been four or five of us in all and it would have been a Saturday close to the start of the summer holidays. I had been intrigued by Farragon for long enough; I saw it from my bedroom window down in Aberfeldy, and though I'm far from being the climbing addict that my father was, I can at least want to know what a particular hill might be like and the view I'll have from the summit. I can't recall now if the McTaggart children had climbed it before but I suspect they had. They were almost made of heather; they could run for seeming miles over the rough and bumpy ground that had me puffed in minutes. They were free-range children who made nothing of what's called bad weather. But they had been born and raised a thousand feet up; all this was their extended garden.

I have no memory now of leaving the farm that day; no memory either of all the planning and poring over the map that must have happened. The first I see of us is as a group, having walked a distance east from the farm and being almost right in front now of the triangle of Farragon. I do know that the ground before us was all being ploughed up for the first time for forestry; going up and over all those ridges left by the machines that did the digging was exhausting. But I remember us stopping too and looking at the various lochs, one of them with a tiny island. Even then, I resolved to come back, for all this was opening a new landscape to me and offering me thoughts of possible new journeys.

But at last we started climbing Farragon proper, and I don't imagine that would have been so arduous, at least not for the young McTaggarts. It was certainly rough walking; a landscape strewn with boulders and smaller rocks, all of it glinting with mica schist. How quite we alighted on the great boulder, I honestly can't recall. The only thing I'm sure of is that for long enough, I knew exactly where it was on the hill. There was a kind of gully, a gash, that ran down from the summit, and it lay close to the bottom of that. I used to sit at my window in Aberfeldy on a clear day and study Farragon and rehearse the walk so I knew I'd be able to find the great boulder again. And remember I definitely did, because I know I found my own way back, not once but several times. If I were to return now, I'd wander about for however long, having long since lost the knowledge; but for years, the location was etched in my head for its importance.

That single boulder was like a rock cake, studded with garnets. I had known garnets well enough in the true Highlands in my mother's childhood bit of country around Drumnadrochit and Loch Ness, but there they were tiny. They had the same distinctive wine-red colour, almost the same shade as a ruby (and of course, garnets have been used and are used for jewellery), but those Highland ones were little more than small round things the size of ball bearings. I had found similar ones at Kenmore too, at the eastern end of Loch Tay. There in the shallow waters of the bay below the church, I found whole handfuls of them.

But the Farragon ones were large as a pinkie nail, many of them at least. I can see me there yet, digging out garnet after garnet from the soft mica schist that surrounded each one. What I remember too is that we were able to clamber up onto the top of that boulder; it was so large three of us could be there at any one time. Centuries of Perthshire rain (and that's a whole lot of water) meant that the garnets had almost been washed out of the schist in which they were embedded. But garnets are a great deal harder than schist; it was so soft you could dig out some of them with no more than patience and a strong fingernail.

Some garnets lay in a soggy mush on the upper face of the boulder, but their beautiful crystal faces hadn't been spoiled or worn. That day, I must have collected hundreds, crouched there finding still bigger and better ones as though some magic wand had been waved and worked. Because the wonderful thing was that this rock cake close to the bottom of Farragon was one of a kind. There simply was nothing else of its size or with such garnets, though certainly there were plenty enough bead-sized ones in smaller stones dotted about on the moorland. As the Australians say, I was like a pig in honey. For however long, I crouched there, digging out one after one. Neither before or after have I stumbled on such a veritable trove of treasure.

Looking for Garnets

The place where rivers meet the Celts believed
was sacred;
they built chapels there and buried precious
things.

One day in mid-July I go down where there is
no path;
breaking my way like a bear.

As the sun flutters
through the green flickering of the trees
I stop and listen, hold my breath,
sweat thick across my forehead and my eyes.

The water is so low it only slides, searches in
among the rocks;
sifting its slow journey to the sea.

No birds, no beasts, only the clear flashes of the
sun;
light dancing in the glades.

And then I break out on the beach,
that place where two rivers meet.

In Search of Gems

I stand there thinking of who was there before
me,
of what might be beneath my feet.

I take my shoes off in case this place is holy;
swirl them in the lukewarm of the water –

and sing something soft, a long time.

Amber

I know for sure I was given an amber pebble as a child; I have it still in my collection of stones. And I know that the person who gave it to me had found it on one of the east coast beaches on Iona. These coves have wonderful names that derive of course from Gaelic: the Gully of Pat's Cow, the Port of the Young Lad's Rock, the Bay at the Back of the Ocean. I'm not revealing which one the amber is supposed to have come from, but the person who found it maintained it was the one location on the island where now and again amber might be found. I myself have searched there, for I know the island's beaches well enough, but I've found none yet. I'll keep on searching.

There's no mention of amber at all in *Scottish Gem Stones* by W.J. McCallien. The lion's share of the locations listed were ones found by the mineral-ogist Matthew Heddle, who would appear to have scurried from one corner of Scotland to another in his passionate search for gemstones. Little surprise that his own collection was of such an extraordinary size and quality. So, even though McCallien's useful book doesn't refer to amber, that doesn't mean it wasn't present here, because I suspect it was.

I think it's worth beginning by thinking of the whole story of amber. The little pebble I was given is probably the size of the nail of my middle finger. If you hold it up to the light it glows, a translucent yellow-orange. It's like a stone made of trapped sunlight. It's fascinating to read just what was thought of amber in classical times. The gem was starting to flood into Rome, and the writings of Pliny, the great Roman author and naturalist, show him remarkably close to the mark as far as working out where it had come from in every sense. His conclusion was that amber had been formed of the sap of a species of pine tree and hardened by frost, heat or the sea. Then it had washed up on the shores of the mainland. Pliny also stated that a collar of amber beads round the neck of an infant would guard them against poison and act as a charm, offering protection from witchcraft and sorcery. In Pliny's time, the price of a small amber figurine was greater than the price of a healthy living slave. That says a very great deal about just how prized it was. A treasure indeed.

As to where amber had originated, Pliny was right again: he believed Northern Europe. Others suggested all manner of places; the more magical the better. But there's something other-worldly about amber; it lends itself to myth. Hyperborea was the name of the magical land it was supposed to hail from, a place that couldn't be reached by land or sea but only by flight. The ultimate north.

Here's the story Pliny was close to guessing. Where the Baltic Sea now lies, there had been an immense forest. Once upon a time, it contained a whole variety of mammals, and birds and insects were every bit as diverse and abundant as today. A syrupy brown resin was released from the trunks of the trees when they were damaged by fires or storms; a resin that disappeared in the fallen pine needles below. As it was buried in the soil, the resin was eventually subjected to immense pressure in a process lasting millions of years. It was like putting a leaf in a book and then piling a whole stack of other books on top. So anything that had fallen into the resin was frozen there forever along with it. And the fossilised tree sap is forty million years old, some of it bearing the remains of the insects from those trees. It's the frozen remains of a lost world never seen by humankind.

After the Romans, the Vikings dominated the sea trade routes in the eighth to tenth centuries. Amber was an important commodity to them; it was used for jewellery, to create gaming pieces and religious artefacts. Pliny was quite right of course about amber washing up on shores. I wonder if he was so intrigued and puzzled by this mysterious frozen stone that he talked to the traders in Rome, asking them to tell him as much as they knew. One wonders just how much of what he and other classical writers thought of amber echoed down the centuries. Its mythic status certainly remained; in many ways it was the myth that grew stronger and the real story of its origins that diminished.

England isn't well off for gemstones generally, but it does have its amber coast that stretches between Felixstowe and Southwold. The beaches that lie between the two are tremendous places to look for amber washed in as the remains of that ancient Baltic forest. But you have to have careful eyes to recognise it; often raw amber looks little more than a dull brown before it's taken and polished to make it the beautiful orange-gold gem we know. And it polishes easily and well.

And so at last we come to the story of amber in Scotland, and in Ireland, for that story is similar in both in ancient times. Amber necklaces from Ireland date somewhere in the region of three thousand years old. Those necklaces are usually found in bogland hoards and in caves. Perhaps they were kept there for safety or as offerings to the river or to water deities. It's surely a sign of the value that was ascribed them that these amber relics are most often found close to gold. This would suggest that amber was a status symbol for the wealthy and powerful. Over 160 beads of different sizes were found in County Offaly in Ireland. Beads of amber and of serpentine were found in a woman's grave on the island of Oronsay in the Inner Hebrides. Four irregular beads of amber were found too in a burial mound over four thousand years old in Aberdeenshire. My questions regarding the amber in both Ireland and Scotland would be how it had come there and where the beads had been made. Had they been traded? Or were they found as treasures on the shore? My guess is that whatever the answers are, and we can't work those out for certain now, it's more than likely the amber carried a mythic status. This must have been seen as a magical stone, with its embedded sunlight promising to bring the wearer protection, perhaps in this world and in the next.

These beliefs in amber as having magical and healing properties lived on down the centuries. They were called *lammer* beads in Scotland and prized particularly on the east coast where amber was seen as a talisman. Perhaps there was something magical in the fact these stones were simply cast up onto the beach, having appeared mysteriously (often after storms). It's more than a guess on my part that amber is found at least now and again on that Scottish east coast (and sometimes on the east coast of islands in the Hebrides too, like my piece of amber supposedly from Iona). Let's remember that in those centuries before the enlightenment, it was a time of superstition. The lives of men and women were hard; a host of dangers threatened their days. Often a deep-seated fear of God was balanced with a strong sense of superstition. A case of belt and braces. Charms and amulets were known across the Highlands; several clans treasured rock crystal talismans that they carried into battle and guarded with extraordinary care for their healing powers.

So it was women on the Scottish east coast wore necklaces of amber beads, and it's my guess they hunted for stones on their own beaches, and that amber was traded. I think we have to deduce that people were fascinated still by where amber came from and how it arrived on the shore. That sense that it had the power to heal seems paramount. In the late eighteenth century, there was a smuggler from Galloway in the south-west of Scotland by the name of Carnochan. He possessed an oval amber bead that he wore around his neck. The story goes that Carnochan removed it from a mound of adders that were in the act of making it. This is how the creation of it is supposed to have happened. The snakes would writhe together at midsummer and produce a bubble in the form of a ring. They would then hiss and blow, throwing the ring into the air with their tails so it would turn hard like glass. It was believed that whoever possessed such a stone would prosper and be able to heal all manner of ills. What I find of particular fascination is that this story is by no means unique; instead, it fits in with all manner of stories told by other cultures in relation to so-called snake stones.

All of this comes back to one central fact: the origins of amber were deeply mysterious. It possessed unusual properties. There was something strange about it that made it different, that gave it special powers. And from earliest pre-history to relatively recent times, that story had remained the same. It was to be valued and it was a talisman. The fact that it was rarely found made it all the more precious; it came from an unknown place and was therefore rendered strange.

In Scotland, this idea of amber being important for eyesight and the healing of eyes is fascinating. An amber stone was kept in safety by one clan for rubbing on the eyelids. In this instance and elsewhere in the Highlands, it was believed that amber would relieve tired eyes and even offer healing to them. Just where had such a notion come from, for it's something that I've read of many times. Was it perhaps as simple as this, that amber possessed light; an amber stone was a cave of gold-yellow light, and by bringing it over the skin the sufferer of sore eyes or diminishing sight was somehow absorbing some of that golden brightness? As so often, would that we might be able to travel back in time to ask them to tell us.

I return to my book *Scottish Gem Stones* and those myriad locations listed by Matthew Heddle. It's my goal to explore as many of them as I can in the years to come and see just how many treasures I can find. And I intend to go back and back to that secret cove on the east coast of Iona in the hope of finding amber. But it's about much more than just the finding. It's about exploring this glen I hadn't known before, or climbing that particular hill I hadn't bothered about before; it's about following certain rivers and going to visit new islands. I feel that with this treasure map, I'm somehow given my country to possess anew. And the emphasis is always on the going there, the being there, the experiencing there.

Heddle doesn't actually specify exact locations; that I mention again for good reason. He offers us clues: he mentions particular islands, definite glens, actual rivers. What haven't ever been found are the maps he used for his countless hunts for gems across Scotland; those would have been marked with specific locations. Yet the paradox is I think that's a good thing. To start with, it means those locations haven't all been visited and dug up and despoiled. It's left us some work to do, and that in itself is a blessing. He's letting us do the searching, the sleuthing. But this way, we'll experience a whole lot more; we'll come back with things a great deal bigger than stones. We'll return with a treasure house of journeys, of experiences, of stories. And for me that's a great deal more precious.

Amber

I dream of going to some east coast beach
one morning in November
when all the sore of winter's blowing in
from Norway and the north.

To be there hunched and small upon the shore
and walk against the wind among the rocks;
then catch in all the grey that shining eye –
one amber bead like sun in frozen stone.

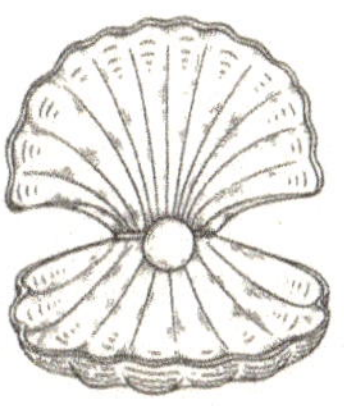

Pearls

When I came to Aberfeldy with my parents in the middle of the 1980s, it was still very much a town on the edge of Highland Scotland. I remember the last Gaelic speaker from Glen Lyon, the longest glen in the country. I remember how she slipped back into her mother tongue over the last weeks of her life, as though starting the journey already to meet her people. The town itself was still locked in quaintness; an emporium and a dressmaker, a post office that hadn't been relegated to the side of a convenience store but where staff took pride in their positions. This was Highland Perthshire indeed. No one thought of locking a door.

When I was sent out during the second half of the summer holidays to pick raspberries and blackcurrants under thundery skies and the air was hot and dancing with midges, I remember the caravans belonging to the travellers among the trees on the other side of the road from the fruit farm. The youngsters worked with us and yet we might have come from different planets: they were maddeningly quick at picking raspberries, just as they were in the autumn at gathering potatoes (or *tattie-howking*, as it was commonly known). They had a fierce pride and doubtless carried within them an awareness of the deep prejudice of the settled world. I thought of it even then, how we youngsters were growing up side by side yet had all but nothing to do with one another.

I remember now that the chief figure in that group of travellers was called Solomon and was, ironically enough, a deeply religious man. I never met him, that I know of, but I carry the memory of his name. And as they camped beside the Tay and the Lyon rivers to pick fruit on the farm, the fathers would be out however often fishing for pearls. I say fathers because it seems to have been unusual for women to be fishers. There would appear to have been strict rules governing the whole culture; by and large, it was men whose privilege pearl fishing was.

It so happens that half the world's freshwater pearls are found in Scotland. It's the case too that the Romans knew this, and that that's one of the reasons they were intrigued to come here in the first place. I like to think that at some point in those early times, a traveller made it to the markets of

Rome. He was curious to follow the rivers and in the end pursued them until he found his way to what was then the capital of the world. And in his purse he had a handful of pearls and he showed them to someone and they wondered, asked questions, even though they had no words in common. But he came back rewarded to his northern land all the same, and what he had brought with him was never forgotten.

It's clear that knowledge of pearls and fishing for pearls has been with the travelling community from time immemorial. A whole culture grew up around it; they created special crooks that could go deep into river water to search for the most likely mussel shells. Because it wasn't that every one might contain a pearl: it was the crooked ones, the misshapen ones. What had happened was that some piece of grit had got inside the shell. The shell almost sensed that this had taken place, that something alien was present. New material, what we know of as mother of pearl, was swirled and smoothed around the foreign body. And more and yet more as the shell grew. So the travellers would keep an eye on the crooked shells that were in one particular river; they would turn them, make sure that all was as it should be when they came back to that river the following summer. It might have been that it was first and foremost the men who had the opportunity of fishing for pearls: the women were too busy at other tasks. One way or another, this was not considered their domain.

I think of the sheer joy that must have been derived from exploring those rivers in the summer months and keeping watch on those mussels that held the pearls. The travellers knew how to lift those particular shells, the mis-shapen ones, and by then another year of growing had passed. It was the joy of summer in nature. As travelling folk they were a part of the natural order of the seasons and the years in Scotland, as much as the Sami people were with their herds of reindeer in Arctic Scandinavia. Yet the paradox was that in both instances the settled community didn't begin to understand their way of life, but instead despised them. The youngsters were mobbed in school and the neighbours in their villages and towns, by contemptuous locals. Yet back then it was the travellers who knew nature best; who were one with the rhythm and the dance of the seasons.

The travellers were able to tell from which river a particular pearl had come, and more than likely those experts from the settled world could too. It was all to do with the minerals that were found in any particular body of water. So there would be a subtle blue sheen from one certain river, or a gold glow from another. I strongly suspect that such knowledge would be handed down from one generation to the next; not only these details, but all the lore that governed this culture.

In the old days, and I'm really imagining the nineteenth century here and the first decades of the twentieth, the travellers would visit the castles and the lodge houses where the well-to-do ladies loved their pearls. On the doorsteps, the travellers would show off their handfuls of pearls, most likely towards the end of those summer months, and be rewarded with a few coins for them. There was a particular jeweller on the banks of the Tay by the name of Cairncross, and the travellers would take their treasures here too, for they knew this was a place they'd get a fair price for their pearls. Until not long ago, the biggest pearl fished from a Scottish river was kept here. You had to ask to see it; the pearl was kept in a safe in a back part of the premises, but would be brought out for you to see.

For long years, my father was a journalist. I don't think he particularly wanted to work for a newspaper, but for better of worse he ended up having to do so. He'd have been happier by far out in the Scottish hills (about which he wrote several books in later years); he was far from content behind a desk in some city office, but he had a family to support and this was what he had to do. In time, he came to write for several magazines in Scotland, chief among them *The Scottish Field*. In those days, it truly was a magazine centred on the outdoors, and consequently my father felt a great deal more at home. I know that for a time he was penning three feature articles each month and under three names. This kind of writing meant he was far more often where he wanted to be, in what I like to term wildscape. He interviewed crofters and island boatmen, climbers and shepherds, weavers and all manner of others besides. One of the many he met and spoke with was Bill Abernethy, and he was a pearl fisher. Not just any pearl fisher: it was he who'd fished the giant pearl, the one kept safe at Cairncross the jewellers in Perth.

But what's interesting about Bill Abernethy is that he wasn't a traveller. He was from the settled world, and I can imagine it was the source of a fair amount of resentment that he should have encroached on their world at all. It makes me wonder all the same if there *had* been always some pearl fishers in Scotland who were from the settled community, who'd learned the tricks of the trade, perhaps having befriended travellers. I know that meeting with Bill Abernethy had a profound impression on my father, and he took a number of pictures of him at work.

Long years after my father had interviewed Bill Abernethy for his feature for *The Scottish Field*, the great pearl was found by this fascinating man. It was very much a double-edged sword. While it won him fame, it did nothing whatsoever for pearl mussels in Scotland. We will never be rid of human greed, and that was what was roused. People had not a clue that to learn to fish for pearls required half a lifetime of learning; it meant knowing how to read a river. More than anything, it meant tending the right mussels for years. So it was that treasure hunters from the settled world went out into the Scottish countryside and found where there were beds of mussels. They ripped them out in their frenzy to get pearls and left the empty shells on river banks. For all but every one would've been empty; perhaps they found the odd seed pearl but it would have been little more than that. These are precious mussel beds that need long years and just the right conditions to grow, and one wonders what may be affecting them now in these days of climate change as we suffer ever more swings of floods and droughts.

Perhaps there was even a silver lining here all the same since pearl fishing became illegal in 1998. It may have been that for a time a few travellers had licences to fish, and that the pearls they found were sold to Cairncross, but by and large, a whole world of tradition had passed into the history books. I know that an elderly traveller who'd been a pearl fisher lived in an upstairs flat beside the bridge over the Moness Burn in Aberfeldy. Even now, I could take you to the door that was his. Would that I'd talked to him and heard his stories.

I went back a few years ago to visit a beach I'd known on the River Lyon. I was with a group of American students and teaching a Scottish course. I wanted them to know about pearl fishing and we happened to be passing this place I'd known of old. I ran with several of them to this beach on the river, for if possible I wanted to bring back at least one shell to show the whole group. The river didn't let me down: there were perhaps half a dozen mussels there on the reddish sand. I took one, even though now you're not even supposed to take a shell if you find one. Some of the largest can grow to be the size of a whole hand; this one wouldn't have been longer than my little finger. It still amazes me that these magnificent shells grow in a river. Never mind whether they have pearls inside; they're treasures just as they are. May we learn to look after them, to keep them safe.

Pearls

They were the reason the Romans came here –
river things, spun into milky globes over years
and years.
I often wonder who it was who found them first;
those mussels, dark shells whorled and folded
like hands in prayer, embedded in deep feet of
shingle.

The travellers knew who they were. The unsettled
people
who followed the seasons, the stars, yearned only
the open road
They carried the knowledge of pearls inside
them, secret,
could tell the very bend over river each pearl had
come from –
this one like the pale globe of Venus at dawn,
this one a skylark's egg, and this the blush of a
young girl's lips.

Yet the Romans never reached the Highland
rivers
where the best pearls slept. They were kept out
by the painted people, the Pictish hordes
bristling on the border like bad weather.

The pearls outlived even the travellers, whose
freedom
was bricked into the big towns long enough ago,
who did not understand any longer the
language of the land.

In the last part of the north,
in the startling blue of the rivers,
the shells still grow. Their pearls are stories
that take a hundred years to tell.

Aquamarine

When I was given *Scottish Gem Stones*, the most exciting thing it contained were locations. Before, I'd been blindfold, hunting wildly in the hope of stumbling over gemstones. But walking with a map is a great deal better than hoping to come across the right locations through luck. And because many, if not most, of the locations listed in the *Scottish Gem Stones* book were found first by Matthew Heddle, when he told you where to find a gemstone, you believed him. Nothing could have been more exciting to me than reading the names of the places he listed for beryl, which is found in crystals, the blue ones having the name aquamarine.

I think this must have been the wellspring for my fascination with faceted stones. There was a relative on my father's side who sailed tea clippers to India. His widow still lived in our home town; he himself was long gone. This sea captain brought back a vast amount from India through the years he served his company. The story was that he'd met an

Indian prince who'd given him an emerald. When his widow died at last at some extraordinary age, my father was made executor of the estate. Huge and heavy trunks were brought to our house and stored in the attic, and I was given permission to go through them. It was the most incredible passage to another world; it was as though I had travelled back a hundred years in time. Inside those trunks were silks and boxes and coins and books. I felt I had stumbled on a cave full of treasure; I was eight years old and a dream had been fulfilled.

And in among all this wonderment was indeed the emerald. Even now as I write this, I remember holding it there in the attic, turning it round in my hand against the light as it sent out its deep green flames, for all the world like some dragon's eye. Of course I begged my father to be allowed to keep it. I worked on him for however long but it was not to be. The emerald was destined for another family member and I had to bid farewell to it in the end. But the memory of that faceted gem was buried in me for ever.

It was through meeting Sri Lankans in Norway that I first discovered aquamarines. That might sound complicated but the story's simple enough. It was at the time of the civil war in Sri Lanka, and Norway had opened its doors to many thousands of Tamil and Singhalese refugees, provided they did not discuss their background in Norway. Some half a dozen students from Sri Lanka were at the college where I was studying. The one thing I *did* know about Sri Lanka was that it was a veritable treasure house of precious stones, and I asked them if they could put me in touch with a seller of gems. One of them gave me the address of a friend and I wrote an old-fashioned letter and sent it off in hope. I waited for a reply and at last it came: yes, he did have a Sri Lankan aquamarine he could sell me. I'll never forget the day I heard my parcel had arrived. I hurried down the mountain path to the village post office to open the box with trembling hands. Inside was the glittering blue faceted stone. So it was I became mad about cut stones, and aquamarines most of all.

It's partly the quality of their blue, and of course they were named after sea water. That link with the sea has been there from the earliest times, and they are one of the oldest gems: in other words, one of the gems we've known and sought and prized the longest. As its popularity grew, so did the myths and legends surrounding it. Aquamarine in classical times was believed to have fallen from the treasure chest of a mermaid. Over the centuries, it came to be known as the sailor's gem, and they wore amulets of aquamarine to protect them from the sea's wrath. It was thought that doing so would calm the sea's rage. Coming into the Christian era, it became the stone of St Thomas, the patron saint of mariners.

Early on, long before the coming of the internet, I saw wonderful pictures of crystals of aquamarine from Brazil and Russia and Sri Lanka itself. And I imagined what it might be like to find a pure crystal of it on a mountainside. There's something extraordinary about a stone so perfectly blue being found in grey slabs of rock; something quite miraculous. That story of how it came to be over the millennia becomes very complex to explain. All of this I had thought about many times before I stumbled on that book and discovered there was aquamarine here in the Highlands of Scotland. Heddle wrote of two locations in particular: the Isle of Arran off the west coast, and a mountain called Beinn a' Bhuird in the Cairngorms. That was the place I set my heart on going to find them. Coming from a hill walking family can have its uses, and I knew it wouldn't take much to persuade my half sister Helen to act as my sherpa, to guide me to the east side of Beinn a' Bhuird.

The Cairngorm mountains are a kind of fortress. There is nowhere else like them in Scotland. You enter their world from one side or another and thereafter are inside them until you withdraw once more. You are leaving behind the mad rush of our usual lives to be there. You become aware of other noises than the ones that dominate our days: you hear streams, the breeze lifting the pine trees, the wind in the high hills once you have climbed into them. It's an elemental place: pure and to all intents and purposes undamaged, undisturbed. I don't believe I would have the courage to go inside the Cairngorms alone. For one thing, I'd be afraid of losing my way, of becoming truly lost, for many have been and still are, year on year. I trusted Helen completely.

So we entered, first through a pine forest. That slow walk in of several miles. Then we were out onto bare moorland and creeping upwards over bouldery ground. A little higher all the time until one by one the sleeping wolves of the mountains that make up the Cairngorms seem to grow from the ground. This is a kingdom now of eagles and merlins and ravens. The only sound is the wind.

At last we're up under the gnarled and rugged head of Beinn a' Bhuird. I can't help feeling excited as a child, thinking of that page in *Scottish Gem Stones* where Heddle's locations for aquamarines are listed. Except now the weather changes. It's summer, but summer can mean nothing in the Cairngorms. This is an Arctic place where the winds can rage with blizzard almost all year round. Now it's great sheets of rain that drive over us as we scrabble foot by foot up the grey corries of Beinn a' Bhuird, but the cold is intense all the same. We're clad for it because Helen has ensured we were prepared for the worst, but I feel that raw gnawing fist of cold that seeps into hands and feet. Now the wind finds us as we climb properly to the heights and there's no more talking and no shared laughter. It's about nothing more than battling onwards, upwards, into driving mists.

I wonder just how many times Heddle climbed here, for he was a member of the Scottish Mountaineering Club and clearly loved being in the hills simply because they were there. But these bens that make up the Cairngorms are vast, sprawling old wrecks of dinosaurs. It's one thing to make a visit as I was doing; something else to know every ridge and face and gully.

Then suddenly we were up at the top of the east side of Beinn a' Bhuird. And it was almost like the lesson from some Biblical parable, because the only thing that lay around my feet were scraggy grey fragments of rock, utterly and completely dull. I might have got down on my hands and knees to start rummaging through them, had I the strength left to do so. But I didn't. I just looked all around at the grey rubble and I suppose I learned something.

I certainly learned something later, however much later, when I found out much more about the gemstones of the Cairngorms. For it's not only aquamarines that are be found here: there's topaz and zircon and great lumps of smoky quartz that's rather naturally given the name cairngorm. The problem is that these gems are ancient. They aren't like the young blue crystals of aquamarine found in the mountains of Pakistan. These Scottish ones have been dragged under glaciers, worn down and scoured and gnarled by the ages. They've been left, somewhere in these grey piles, washed out remnants of the memories of gemstones.

I'm rather afraid all of this may read like something of an anticlimax. For of course, in many ways it was one to me that day since I'd dared to believe in my naivety I'd come home with a blue gem from Beinn a' Bhuird. But what that day taught me was something worthwhile all the same. It taught me to see a day of treasure hunting for gemstones as experience in parallel. In other words, you don't cast everything else to one side in your driven desperation to find the one gem, so that if that doesn't happen the expedition is a disaster. Somehow, you work to take every piece of the day and experience it to the full; you may have a goal of a gem at its end, but the whole event is no longer doomed if that's not found, for you have found everything else that mattered too.

But I don't deny I still dream of finding the blue crystal, the aquamarine, however small it may be, one day. My half sister Helen died far too few years after that day of wild walking in the Cairngorms. I didn't see her at the time of her passing, nor had I had the chance to do so in the preceding few months. She left me just one thing in her will: the most beautiful royal blue velvet drawstring bag with a Celtic cross stitched on one side in white. She left no note of explanation with it but my guess is she meant it as a bag for treasured gems, and that's what it's become.

Aquamarine

Listen. Here there is nothing but everything.
Look down into the water that has lived here
since the beginning. By some miracle
flowers the size of a child's smallest fingernail
cower out of the gale. Imagine
what it might be like to climb here one winter night
as the full moon edges silver-coined
from out of blue-black empty sky.
To wait here through pure night
and catch a glimpse of falcon folding into fall
from that cliff edge lit, smoothed sheer.
Has anyone in all the world been here before;
this place where now you wait for dawn?
And so it comes, a black blood in the east
and then a glowing that becomes at last
the beginning of what might be morning.
And as you turn, the cold goring hands and feet
you see somewhere inside the gravel's grey
one single see-through piece of blue
as clear as sky or sea. You cannot quite believe
it's real, just watch it dumb then stagger round
on hands and knees and dig and dig
until it's yours. Then all the sunlight there could be
bursts into flame and fills the crystal full.
It's yours, a blue gift that waited here for you.

Smoky Quartz

I seldom found anything other than serpentine on those summer days in the Hebrides. The beaches offered shells and I contented myself with those: cowries and the blue goose barnacles attached to old spars of wood. I longed to find an intact sea urchin for they must be the great joy of any beach, but these fragile jewellery boxes with their red and gold and amethyst markings all too seldom survive the strength of most Hebridean waves. What you *do* find are their remains: fragments of sea urchin shell no bigger than a thumbnail. Only once have I found a whole unbroken globe of sea urchin. We were in a corner of the west of the island of Mull and the beach closest to our cottage was sheltered enough; I recall it now as muddy and rather undramatic but full of rock pools. And there in one of them I found a sea urchin that seemed almost to have taken on the colours of the beach. It was the smallest I'd seen in my whole life, for I had a collection at home, all from the Scottish islands. This one I almost passed by, so grey it was and dull.

But the longing to find stones and gems never left me for long. I remember visiting the island of Eigg with Douglas, a friend from school; my parents had found a caravan on the west side and we stayed there. I remember yet the views we had out to the great tented island of Rum, the dark sheets of its mountains descending to the sea. Douglas and I would have been twelve years old; each new day was long and composed of exploring. It was that longing to find treasure; you were old enough to know it almost certainly would never happen, yet the child within still believed it might. We weren't far from a beach called the Singing Sands: when you walk over the wet sand your feet cause a kind of rubbing so a sound is made for all the world like some kind of singing. At any rate, one morning Douglas and I were on one or other of the beaches there on the island's west. We were looking as always for anything exciting that might have washed in overnight, and suddenly we stumbled over a whole consignment of teabags. Hundreds of drenched and stinking bags, washed out of the crate that had carried them, but that wasn't enough to send me on my way. Still there might be something to find.

And there was. In the middle of that slimy pile of teabags was something else. A tiny cotton bag, not much larger than one of those special stamps made for some royal occasion. It had been sewn neatly round the edges with white threads and the bag was full of something – of what felt like small stones. All I remember next is running back to the caravan over the beach as fast as ever possible, then my mother taking scissors to snip one corner of the tiny bag and spilling it out onto a saucer. Minute white stones, perhaps eighty or a hundred in all. They had a milkiness about them, a sheen; almost resembled tiny fragments of teeth. But I knew my stones well enough to recognise them at once as opals. My mother disagreed vehemently; she was sure they were nothing more than some kind of substance intended to keep dry the consignment of tea. She insisted on putting the new-found gems into boiling water, assuring me they would dissolve. I waited nervously, afraid I might have been wrong. But they didn't dissolve: they dried unchanged, and triumphantly I knew still they were opals. I have the little bag yet and whatever their true story might have been, they remain the closest I've come to treasure. It's a story of discovery you almost feel was wished into happening.

But apart from that tiny bag of opals, there wasn't in truth much I did find as a gem hunter on those Hebridean beaches and west coast shores. There were occasional pieces of smooth and shining flint, most often the colour of toffee. There might even now and again be fragments of agate, especially on the beaches of the island of Mull. I think I learned to content myself with *not* finding exciting stones on those west coast holidays; I knew that at some point or other, we'd be back on Iona and going down as ever to St Columba's Bay. It was enough to wait.

The rest of the summer was made up of the weeks back in Highland Perthshire, and for much of that time I was out picking fruit. I battled with July and August for always they seemed warm and wet. The air muggy and thundery; the skies grey and airless, the river thin and a shadow of all it had been. There were occasional forays into the hills with friends and there were the days of finding those gorgeous pools that still remained deep, somewhere in a wood and under the tail of a waterfall. The days dragged, and I felt listless and strange. I wanted the summer over, for the best of it was gone. Now I longed for autumn.

And almost always in September, we'd be in Strathspey for a week, under the shadows of the 'sleeping wolves' – the Cairngorm hills. If I think of it now, I can catch the scent of being out in the early mornings. I consider it now a little Norway, for the parallels are so many. There'd be birch woods, some of the trees ancient and half-fallen, their boughs studded with bracket fungus. I remember going and breaking those pieces of bracket fungus from the old roots and fallen branches; always, I'd take them and bring them to my face and smell them, and they made me think of some kind of bread. But the woods had the quintessential scent of autumn; of blaeberry and cranberry patches, of deep moss, of juniper bushes, of lichen. That was sufficient to thrill the heart because summer was past and the days were getting shorter; all that I loved about the year was ahead, was in waiting. Almost always, there seemed the smell of wood smoke in Strathspey too, for the nights were colder and there were full skies of stars once more.

I've always loved that freedom of walking off into a wood. Here a path was seldom needed because the woods so often were open; a few birches here, a gnarled pine there, a rowan tree somewhere further off. There would be a carpet of deep grass or heather; many times, you were able to look deeper, further on, because of the thinness of it all. And a river would never be far away.

Smoky Quartz

By this time in early autumn, the rain would have returned in earnest and that meant the rivers or burns were strong things, and always the colour of whisky, or darker still. In fact, they were the colour of the best cairngorm, smoky quartz. The very finest boulders or chunks of cairngorm rock, named of course after the hills, are all but black. But those are rare and seldom found. Since the stone is just a variant of quartz, it's common enough here; the best places to find it being river banks or the summits of the hills. Most of the pieces I found by those burns weren't black at all: they ranged from a dark orange to gunmetal grey.

There's something about finding crystals, perhaps embedded deep in boulders of quartz, that's hard to beat. I remember years ago, when staying in Ballater, on the banks of the River Dee, discovering crystals in among the great beaches of rock debris up from the river's flow. To reach the crystals those boulders had to be broken. I tottered with them in my arms and bashed them against other rocks until they burst open at last, for there were fissures in their edges. I would hurl the boulder in such a way that those cracks and crevices were the part that met the other rock. Those wonderful moments then when the one I carried shattered and the crystals were revealed.

I've read in more recent years of the extraordinary cairngorm treasures that were brought back in Victorian days from the bigger rivers of Strathspey. For the stone has been well-named: this is very much its heartland in Highland Scotland. Both women and men were going deep into the wildscape of that moor and hill country to find and bring back whole boulders of black quartz. It was and is used by jewellers; I associate it with Highland dress, with the cut stones for kilt pins and the brooches for ladies' tartan skirts.

Only once do I remember finding a truly black piece of cairngorm, and I have it yet. You had to be careful hunting for the finest chunks because often it meant breaking a large boulder to reach the cairngorm locked on one edge. The shattering part wasn't the difficulty – that happened easily enough – but what you had to watch out for were the fragments that flew in every direction. Sometimes, of course, the precious part of the rock you were trying to reach would be ruined; that was the very bit that would break. But then you had to be mighty careful searching among all the fragments for the darkest pieces of cairngorm: they were as jagged and sharp as shattered glass. You learned from experience, but such was my keenness and my addiction to hunting for gems that never once did I think of wearing gloves or protective glasses. Fortunately my eyes and hands are still here, pretty much none the worse.

I know for absolute sure that some mighty cairngorm treasures must remain on the banks of Strathspey rivers. The gem isn't prized as once it was: it has somehow just returned to being deemed another variant of quartz. There's fashion in regard to gemstones too: they've always come and gone in terms of value and desirability. Well, some do at least; others never seem to lose their sparkle and allure. All I know once again is that many parts make the whole; those early autumn days by the rivers were magical. Time lost all meaning; I had the scents of a hundred things in nature around me and the Cairngorm hills above me. I was in the middle of incredible riches.

Cairngorm

I'd climb all day until I reached the roof –
a granite plateau made of moss and snow.
I'd bring no water, rather stretch down deep
through pools of clear and freezing cold to find
and fill my thirst. I'd pitch my tent and wait
until the full moon rose on midnight, turned
each shard and fleck of stone to silver-white,
as deer clicked out across the brittle rim
so not another thing might be alive
until the very edges of the sky.
I'd waken, strange, and see the newborn sun
had made my face pure amber in the dawn.

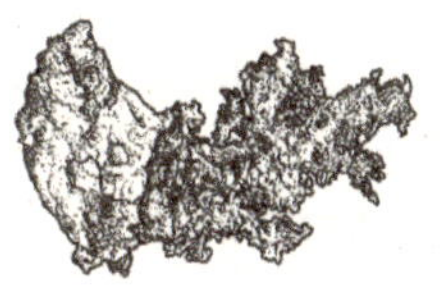

Gold

Like many another child, I was born a treasure hunter. As my wife puts it rather unkindly, I'm a science-free zone. I could never have had the patience or the interest to become a geologist , but I did learn a fair amount as a magpie all the same. I know, for example, all about hunting for cubes of iron pyrites (known more popularly as fools' gold) that glistened in slabs of slate. All someone had to do when I was perhaps twelve years of age was to give me a hammer and I'd be content for hours, hunched like Gollum in his cave, finding new pieces of *precious*.

But in those days I didn't know that I was growing up in a country particularly blessed with exciting gems. Perhaps it's partly because Scotland has such a quantity of high hills; perhaps the number of old volcanoes has something to do with it too. But all that really happened in those early days of my hunting was that I stumbled on things. I had a good eye for stones (partly because those eyes were constantly searching the ground around me), but of course a great deal of it was about luck.

Often the locations listed in *Scottish Gem Stones* had been found by the rather extraordinary and doubtless eccentric Matthew Heddle. Heddle, who like almost everyone in the Orkneys had Norse ancestry, was born on the island of Hoy. I like to think of Hoy as the Highlands of Orkney; it's the one part with serious hills and glens, a sense of wild-scape. Heddle was a collector: when he went from Orkney to school in Edinburgh, he started putting together a herbarium, but rather fortunately for us, this was dropped by accident into a stream by one of his friends. Heddle wanted to start collecting something else so he decided on geological specimens. After taking different paths, and for one reason or another, Heddle eventually devoted himself to the collection and study of Scottish minerals. And by this time, he was very much a mineralogist; a great deal more than just a magpie. The result of all his devoted collection became the Heddle Collection of Scottish Minerals, which now belongs to the nation and is considered one of the finest individual mineral collections that has ever been made – not my words but rather those of W.J. McCallien.

So much of Heddle's life must have been taken up with searching the remotest parts of Scotland, it's remarkable he had the number of children he did or any kind of home life, yet somehow or other he did.

One apparent discovery by Heddle has become of particular significance. The northernmost Munro in Scotland is Ben Hope in Sutherland, and Heddle thought he'd found evidence of diamonds just to the north of the Ben. The trouble may have been that Heddle came back to Scotland to begin his rock-hounding in earnest having just experienced the diamond mines of South Africa. There may still have been something of a glittering in his eyes, as it were. It's clear that Heddle made very few mistakes when it came to identifying gems, but there seems to have been a bit of wishful thinking here. That having been said, it appears that the possibility of there being diamonds in the hills of the northern Highlands hasn't been discounted entirely.

Scottish Gem Stones is definitely a book for magpies. At once, I started copying down the locations for amethyst and smoky quartz and onyx and whatever else. I began dreaming of visiting these places and coming back laden with treasures. Except it isn't quite as easy as that. Heddle's locations are seldom exact. He might mention a particular glen for the finding of a certain gem – let's say tourmaline – but have you ever ransacked an entire Scottish glen in the hope of stumbling on that one stone? And here the weather has to get a mention. Just try looking for that rock (and I have done it) on a howling day in October when the weather is coming at you from every direction at once. If the search involves looking in water (which frequently it can do), then try having your arms plunged up to the elbows in what the Scots call a burn for more than ten minutes, even in the month of June. And we haven't even mentioned the midges. The more I read about Heddle, the more I admire the obvious tenacity he possessed for his hunting.

But then I met a living hunter of Scottish gems. He happened to have his home close to where my mother had grown up in Glenurquhart. This man, who was called Sutherland, lived a thousand feet above the loch, and most of his house was devoted to stones. Like me, he was a Gollum. And like Heddle, he did a vast amount of walking and climbing to reach the right locations. I can still remember the stories he carried with him too.

The most exciting of those had to do with gold. Not many years before, he had been travelling with a group of geology students. They were somewhere in the Highlands with their minibus and had got onto the subject of gold. He suddenly realised he had an old gold pan with him and could take them down to the river they were passing and show them how it was done. They wouldn't actually find anything, but he'd at least show them. Down they duly went, and he scooped up a first lot of silt from the river. He swirled it in the water and there in the middle of all the greyness was a huge nugget of gold.

The trouble is that all that glisters is not gold. I did know in my childhood there was gold in corners of the Scottish Highlands, and consequently I believed that all the glistening specks on the beds of river pools must be it. Of course they weren't; they were instead tiny flakes of mica gleaming in the sunlight. There are specific locations that might have bits and pieces of gold, but you might spend an entire lifetime looking, having missed the one speck half a mile downstream from you.

That said, Highland Scotland did once have a gold rush, short-lived thought it was. And it was in Sutherland, one of the northernmost counties of the mainland. What happened was this: a man by the name of Robert Gilchrist found fragments of gold in two burns, the Kildonan and the Suisgill, in 1868. These were hard times in the Highlands: poverty was rife and many of the glens had lost those who had lived there from time immemorial through the Clearances. And here in this very corner of Sutherland, those Clearances had been the most brutal; families were burned from their homes. The Duke of Sutherland was a man whose name would be remembered for all the wrong reasons.

But so it was that word reached the newspapers of the gold that had been found by Robert Gilchrist. Some 600 prospectors found their way to the Strath of Kildonan in the spring of 1869, hoping to make their fortunes. The picture of their shanty town looks like any from the many there have been in countless corners of the world, except around this one the bare moorlands stretch away under huge Highland skies. What a harsh year it must have been, living in those wooden shacks high in a wildscape of heather and bare rock, coming back drenched and freezing in the evenings, having found nothing in the deep peaty water. It's clear the Duke of Sutherland didn't like having people back on his land; he preferred sheep. First, he decided that gold mining licences would have to be bought: £1 per miner per month, which was a vast amount in 1869. More was to be paid by anyone lucky enough to find any actual gold. Consequently, most of the men drifted away, tempted back to the greater certainty of fishing or working the land, if such things were possible to them at all in Sutherland now. Then the Duke decided enough was enough: this gold rush fever was to cease from the first day of 1870. What he saw was that all this frenzy of activity around two of the streams in his vast estate was threatening to interfere with the much greater revenue he could derive from fishing, stalking and grazing. He was good at getting people off his land. The truth is, all the same, that the gold rush might not have lasted much longer anyway. Probably Robert Gilchrist had simply been lucky.

Of course, the finding of gold must be a fabulous thing in purely aesthetic terms. Gold doesn't rust: it remains the same as it was from the beginning. Little wonder it became the symbol of kings and queens; this strange and almost magical thing at the bottom of river pools and hidden deep in mountains. I have to say I hope there isn't ever another gold rush in the Highlands. For me, the locations are far more important than the gold, and sadly we human beings have an awful habit of leaving those locations a lot worse off in our madness to reach the elusive treasure. I think Heddle would have agreed. It's clear he loved the hunting of his gems, but as much because he loved the wild places themselves as anything else. His passion was for exploring, for experiencing the magic of the light, the water, the hills. He came back with extraordinary things, but he carried them out of extraordinary places.

All That Glisters

There is gold here to find
but not the kind you thought;
not that dreamed-of hoard.
Here in the amphitheatre of the hills
where the sides of the sky are torn
by the tugging of the guy-ropes of the wind
and the lochans ruffle in their blue-grey pools
a thousand feet above
the rush and hum of boring roads
just now and then the sun breaks through
to set alight some nowhere place
right in the middle of everywhere
and for a moment purest gold is yours
to hold, to keep, forever.

Agate

Perhaps Heddle's favourite stone of all was the agate. There are lots of different types of agate and plenty enough famous agate locations around the world, and Scotland is blessed with a fair number. There are good sites in Robert Burns country in Ayrshire, the country's south-west, but by far and away the best are in the east. By that I mean the counties of Angus and Fife, both the shores and inland too. I'll talk about the inland localities later, but first I should try to offer a sense of what agate is for those for whom it is nothing more than a word.

As mentioned, there are lots of types of agate, but it's what's called banded agate I'm writing about here. Often slices of banded agate are polished; you may have seen a slice hanging in the window of a gem shop with all its gorgeous coloured rings. I always think of them as being akin to the rings of a fossilised tree. Concentric circles, as though revealing the age of the stone. The wonderful thing about banded agate is that each one is unique; you'll never know from looking at the outside what the heart of it may reveal. Sometimes that can prove a great disappointment; it's like seeing a fabulous caterpillar with the most vibrant colours and believing for sure it will turn into the finest of butterflies. To your dismay, it becomes instead something little more than a dull white. The converse can be true with banded agates, and perhaps with butterflies too; a rather dull exterior can be hiding the most wonderful inner landscape of rings and colours.

There's a lighthouse at a place called Scurdie Ness on the coast of the county of Angus. Years ago, when first I heard whisperings of these sites for agate but long before learning of Heddle and his listings of gem locations, I battled my way from the railway station at Montrose to the Scurdie Ness lighthouse. I say battled because never will I forget that cold. To the north, the Angus hills were powdered with snow. Geese flew in skeins over the bird reserve at Montrose Basin, and my hands were plunged as deep into my pockets as possible because of the razor knife of the wind. It came straight against me from Norway, such cold I'd have struggled to speak had I needed to. By the time I was out to the Scurdie Ness lighthouse and its straggling shores and rocks, I confess I hardly cared about agate any more. All I wanted was tea and warmth; my whole face felt frozen by the cold.

But I had come this far. At the little stretches of shore, I bent down out of the wind to start searching, and it wasn't long before to my joy and amazement I'd found fragments of agate. Often they were polished, coloured toffee and gold. Then I found a piece of amethyst the size of my pinkie nail, the loveliest deep mauve. There were crystals too; some broken out of bigger rocks and others still held in them. I didn't forget the cold entirely but nor could I tear myself away all the same. That was the first of many visits I would make over the coming years, and for whatever reason it always seemed to be that I was there in early spring. There truly is no cold like the cold that comes from the North Sea.

So I imagine Heddle here a hundred years before, very close to Scurdie Ness lighthouse. One of the most famous agate locations lies just half a mile south, on the coast, at what was the settlement of Usan. It's an ancient place, dating from at least the thirteenth century, but perhaps a great deal older still. Of course, like many other settlements dotted along the coast, it was built on fishing. And it was a wonderful source of great agates, the reason it attracted Heddle.

It's said that in the late nineteenth century, so possibly after the time of Heddle's many visits, a solitary collector, a hermit, occupied a small cottage just above the high-water mark on the Usan foreshore, making a living from extracting stones for visitors. Perhaps Heddle *did* encounter him; perhaps they talked together about agates and shared their stories. Perhaps Heddle haggled with him for his very best treasures. What's certainly the case is that the ruins of the hermit's house are visible to this day. Usan was the site of the so-called Blue Hole agates. It's a name that has gone down in history and acquired iconic status, because the precise location of the Blue Hole has been lost. Doubtless it would have been scribbled on one of Heddle's maps. Have a look online at some of the banded agates from the Blue Hole. I think that the loveliest of them have a colour akin to the fur of a Russian Blue cat, the most beautiful smoky melding of grey and blue.

So eager was Heddle to get the best specimens, he used dynamite. If he did encounter the hermit, I wonder what the man made of that. It smacked rather of addiction, but then Heddle wouldn't ever have had all the time he wanted at his beloved agate locations. He had to be moving on, searching for sites for jasper and fluorite and carnelian and whatever else. He had a whole country to map.

Agate

At some point in past years, I met a couple who were agate collectors. Once upon a time, Robin had been a composer, but the music he'd written hadn't met with real success. He and his wife Jean had had a family; for years, they'd lived in a corner of northern England. Since the children had flown the nest, their living had come from finding, cutting and selling agates. Each spring, they drove up to the Scottish east coast, staying in Fife or Angus for perhaps a fortnight, then drove back home with a whole tonne of agates in the boot of the car. Yes: a whole tonne.

Very quickly, they got to know of my puppyish delight for gems in general and agates in particular, and they invited me to join them for a day when next they were in Scotland. I discovered over the times I hunted with them that they were by no means alone in operating like this; there were a good number of other hunters too, but Robin and Jean were perhaps alone in actually making a living from agates.

Fife and Angus aren't mountainous: both are composed of rich farmland. They aren't exactly flat either; I'd describe them as bumpy. A sudden ridge of hillside here, a small bit of woodland there, then fields with umber-coloured soil stretching to the valley below. Landscapes ribboned with single-track roads and farmhouses planted now and again in what I'd call the middle of everywhere. Good, strong, stone farmhouses built to withstand everything the weather flings at them from the west. That was where Robin and Jean had started: they'd got to know the farmers at places with names like Gourdie and Pitroddie Den, Tinkletop, the Path of Condie and Ballindean. As I write them now and listen to their sounds, I think they might have been conjured by Tolkien and from out of The Shire itself.

Robin and Jean would write and ask a handful of these farmers if they could come to hunt for agates. It was important to time the visit correctly; it had to be as soon as possible after ploughing had taken place. Any later and heavy rain would make the rocks claggy, which would mean it was hard to look at them properly. They told me most of the farmers shrugged their shoulders and said yes; hunting for agates was as meaningless to them as looking for kangaroos. The following year, Robin and Jean liked to take a slice of polished agate as a gift to each farm where they'd searched for stones, just to show them what was buried in their fields.

After meeting at the station in Dundee, we'd trundle out in their old van to some corner of Fife or Angus. It was early spring and often there'd be snow. We'd park by a farmhouse and go out into the field, dressed as though for Siberia because of the intensity of that cold. For hours, we trailed up and down the lines the plough had left, heads bent to catch any glimpse of what might be an agate. You had to get your eye in; learn the look of one. To start with, they're about the size and shape of a potato. But there's a skin to the outside of an agate you get to recognise that's hard to describe. Something that for me makes them dragon eggs; unlike anything else. Sometimes there's a strange eye in the agate's skin offering a tiny window into the world within: an orange-red or a silvery-blue, depending on the shade of the core. And so, from perhaps ten in the morning till five in the evening, the three of us wandered the field. Sometimes the cold got too much, for once I remember a whole day of snow flurries. Then we huddled behind the van, sipping tea and showing off stones.

The agates I'd found would be kept in a separate box before being taken back to the north of England. I waited in Perthshire for all the world like a child, wondering when my box might arrive. For Robin had a diamond saw; he cut the stones first and then polished the resulting faces of the agates. Often the ones from Fife and Angus have the colour of pantiles; sometimes they can be reddish and sometimes that fabulous blue-grey I've mentioned. And, as I've said, each one is unique. Many of the banded agates have a patterning that's almost like a sky full of the flames of the Northern Lights.

The loveliest agate I've ever found was at the end of a day's hunting with Robin and Jean as the light was dwindling and we were making ready to go. Just as we left the field, my eyes alighted on a last stone. I picked it up, somehow knowing what must lie inside. When Robin cut and polished it and sent it to me, I recognised it at once. The most fabulous light grey sheen to it, and rings in the shape of a kite. So of course that's what I called it.

Once I was visiting Lake Superior in the Upper Peninsula of Michigan. I discovered there were agates to be found and of course I combed the shores in the freezing wind, remembering those days of hunting in Fife and Angus, my head down and searching as of old. I found not a single one, but later that day I visited the home of an agate collector who lived in the tiny settlement close by. She brought out boxes of agates to show me and we were united by our love of these extraordinary creations and by our stories.

Scurdie Ness

Where the geese rise in long
slow arrowheads, where the winter sun
strikes like flint at the low hills.
Where the sea scars in from Norway,
sharp and colder than snow.

All afternoon we crabbed about that headland
looking for agates, searching through stones
for little bits of pink and white with rings.
Out on the edge of the sea a ship fought south;
the cormorants gathered in covens on the rocks.

We came back against the wind, hands raw
like fleshless bone. The sun sank into flames of
cloud
and in a shudder the whole of Fife
was wintered. Our pockets rattled with knuckles
of agate
as the first snow petalled the hills.

In Grand Marais

The biggest lake in the world. May.
We hunch the shore as driven waves
card the sand and stones. No agates,
only hands left red and raw, held clenched in pockets.
The car hums a back road out of Grand Marais
and all at once signs for a museum, an agate house.
It looks asleep, its windows shut like eyes,
but then a shadow beckoning us round the back.
We slush the grass, curve past
a boat taller than us both, once blue, left chipped and dry.
The woman's on the step, all smudged with paint;
she's busy, could we come back tomorrow?
No. Sorry. She reads the disappointment in my eyes,
vanishes, returns a moment later bearing
an old wooden box with a single leather strap.
I sit beside her on the doorstep;
she brings out agates from Mexico, Brazil and China,
from every secret cavern in the world:
dragon eyes – flame-ringed, translucent, cut and loved –

shining like strange and priceless moons.
I sit on that step in Grand Marais
and catch sometimes the rough weather of her face,
and think of her journey, the places she was beached
before being washed up here, on the shores of Lake Superior,
where winter's six feet deep and summer's boarded up
by late September. And I think too, as I sit beside her,
sharing stones and stories, four thousand miles from home –
this is what being human means so much.

Diamond

Diamond is what results when carbon is put under tremendous pressure. Coal is a stage in that journey; it's for good reason it has been called black diamonds for long enough. Today, it's possible to create diamonds; the necessary pressure can be achieved artificially. We think of diamonds classically as being white, but some of the most famous diamonds are of other colours.

There's an intriguing treasure tale to be told at this story's close, even though it has perhaps a rather frustrating conclusion. Matthew Heddle went out to South Africa to report on the gold mines that were being developed there, though we don't know precisely when. What we do know is that the Kimberley diamond fields were discovered in 1870, and immediately one wonders if he visited them while there. But he and others obviously hoped to find diamonds in Scotland; there were extensive searches undertaken in Fife especially, though evidently without success. Heddle would have been well aware that a well-known Ayrshire geologist, immemoriably bearing the name John Smith, had found microscopic crystals of diamond in Scotland. So the desire to find diamonds would have been there, and perhaps the hope, if not the belief, that they *could* be there. John Smith's find was in his home county of Ayrshire.

Nor do we know either when Heddle climbed Ben Hope; perhaps he explored it a number of times and became interested in that district particularly. At any rate, Heddle found a stone, as he described it, *near a lake three miles north-east of Ben Hope.* Anyone knowing this area at all will realise this was akin to describing it as *somewhere in a patch of grass in a hay field.* It was clear Heddle knew exactly what he meant and doubtless would have been able to take someone to the location in the blink of an eye. Hardly so easy for anyone else. The specimen he discovered was described intriguingly as containing either *colourless garnets or diamonds.* This was in 1901.

Then the specimen went missing – for the greater part of the twentieth century. There was plenty interest in Heddle's diamond, naturally enough, but self-evidently it was vital to be able to see what he'd found back in 1901 to know what was being looked for now. That area to the north of Ben Hope is made of moorland studded with myriad lochans. Those who visited in the hope of stumbling over further diamonds found no evidence of any sizeable loch.

The search for the missing specimen didn't stop, because of the interest in Heddle's claim, since he'd been such an eminent mineralogist. Alexander Thoms, his son-in-law, married to his daughter Clementine, had been a collector too. His collection wasn't known and famed to the degree of Heddle's, but it survived all the same and was housed in Dundee. In the early 1990s, it was transferred to Glasgow, and in the fullness of time to the renowned Hunterian Museum. It was there that the mineralogist Dr John Faithfull began to consider the story and wonder too what might have become of Heddle's diamond. He himself had been interested in Scottish minerals and gems since his childhood days and was familiar enough with the story of the specimen and with the names both of Heddle and his son-in-law.

Thoms's collection was in excellent order and accompanied by a numbered catalogue. The second entry was for a specimen from Ben Hope that might have been a diamond, but though almost all the other specimens were located easily enough, the second one wasn't there! It was almost as though Heddle had arranged this elaborate and mischievous treasure hunt at the close of his life in order to madden and perplex his successors. But then Dr Faithfull had occasion to visit one of his colleagues; he happened to catch sight of something and recognised at once the handwriting of Matthew Heddle! And it was nothing other than the missing specimen, which after close to a hundred years could be examined at last to find out if it was colourless garnet or diamond, as Heddle himself had wondered. The rather disappointing and frustrating truth is that it was the former; there'd been somewhat too much wishful thinking back in 1901.

And yet it may be, paradoxically enough, that he wasn't entirely wrong either. There have been further searches for diamonds in the Highlands, not least because they've been discovered in other northern areas (principally in Canada and in Russia, and in abundance). Diamonds have been found both in the north of Norway and Finland; those discovered in the northern mountains of Sweden thus far have been very small. Back in Scotland in 1995, a very large and valuable sapphire was found on the Isle of Lewis, in the Outer Hebrides.

So perhaps Heddle was on the right trail but just at the wrong point. I like to think that he would have been highly amused at the long hunt he set in motion. Reading of his life, it's clear he was no dry academic interested only in his science labs and what he discovered under the microscope; he had a passion for his country and for exploring every wild yard of it.

He left us not only with a unique and priceless collection of Scottish minerals and gems that's highly unlikely to be matched, let alone bettered, but he left us with a countless number of journeys to replicate. I think that's what he would have wanted.

Dreaming

There are no diamonds in Scotland;
not glistening and white and big –
but as a child I hoped that there might be,
that in the dreich days of November
when we tramped the scraggy hillsides
my eyes on granite fragments
worn down by glacier paths –

I just might find one single see-through gem
that was.

Appendix: On gem collecting

Across the millennia, fairy tales, stories, songs and cave art – not to mention archaeological records – tell us that precious stones, gems and shells have attracted humans from the earliest of times. Whether worn as adornments, presented as offerings to powerful rulers or vengeful gods, hoarded for wealth, or buried with loved ones, it's abundantly clear that these riches have always been prized.

Jade was already crafted for use in tools and adornments in China by the time the Neolithic era began, some seven thousand years BCE. We can see from grave goods and hieroglyphic texts how treasures like lapis lazuli and amethyst were worn and worked by the ancient Egyptians. Sanskrit texts described early systems for classification and qualitative assessment of pearls, diamonds and other gems, for trading purposes; and, later, an ancient Greek treatise documenting gems and minerals, *On Stones* (c.300 BCE), was written by Theophrastus, a pupil of Aristotle. Some three hundred years after that, Pliny the Elder catalogued many gemstones in his *Natural History*.

Gradually, our growing understanding and practice of scientific study has led to more sophisticated classification of minerals, through visual, chemical and physical testing, crystallography, spectroscopy and more. We can measure hardness, brilliance, colour, clarity, lustre and other metrics. We can create synthetic jewels with often similar properties. Yet such analysis does not begin to account for our strong desire to collect, give and adorn ourselves with gems, minerals and precious stones.

Gems are not just valued for their objective 'worth'. They are truly precious because they possess a magnificent combination of beauty, rarity and (usually) durability.

Gemstones may be either mineral or organic in origin. Mineral gems are crystalline compounds with a defined chemical composition and structure. Examples include relatively common crystals such as quartz, and scarcer ones such as diamonds.

By contrast, organic gems are the result of living organisms or geological processes transforming organic substances, often gradually and over long periods of time. This type of gem includes amber, which is a fossilised tree resin; coral, the skeletal structures of marine invertebrates; jet, a form of decayed wood; and pearls, made by layers of nacre building in oysters and other shells.

Another classification of gems is precious metals, naturally occurring and usually rare. These are softer than their crystalline, stone and (often) organic counterparts, and can be worked by shaping and engraving to forms jewels, containers and ornaments.

There is no worthwhile distinction between precious and semiprecious gems and stones; these arbitrary labels don't reflect their rarity or beauty. In general, the hardest minerals – diamonds, rubies, sapphires and emeralds – have been the most consistently prized, but softer stones such as topaz, tourmaline and opals have been just as coveted in various cultures. Their desirability is, of course, largely in the eye of the beholder.

Across the globe, there are 'hotspots' where rich gemstone deposits occur, including Myanmar and Madagascar, where colourful sapphires, rubies and emeralds are found, and Russia and Botswana, which both have vast diamond deposits.

Scotland's most abundant gemstones include agates, quartz (especially Cairngorm and amethysts) and garnets, as described in this book. Other distinctive northern European gems include Scandinavian nephrite jade and England's Whitby jet.

Scientific study and various practices have by now described and categorised innumerable valuable gemstones, but there is little to match the thrilling experience of finding beautiful shells, crystals, stones, fossils and gems in natural surroundings.

To learn more about the geological processes that formed them, and the classifications and uses of these stones, there is a very brief list of introductory sources below. Many people also attribute healing and spiritual properties to crystals and stones, and there are plenty of sources available with inspiration to pursue those subjects, whether for meditation, talismanic or similar uses.

If you go on your own forays for gems, always be mindful of the environment and wildlife, and remember that rare stones will likely be subject to local regulations prohibiting their removal from wherever you find them.

Happy hunting!

Deer, William Alexander, Robert Andrew Howie and Jack Zussman, *An Introduction to the Rock-Forming Minerals* (Mineralogical Society of Great Britain & Ireland, 2013).

Heddle, Matthew, and John George Goodchild, *The Mineralogy of Scotland* (Hutson Street Press, 2025 edition).

Marshak, Stephen, *Essentials of Geology* (W.W. Norton, 2009)

McCallien, W. J., *Scottish Gem Stones* (Dabney Press, 2015)

Nesse, William D., *Introduction to Mineralogy* (Oxford University Press, 1999)

About the author

Kenneth Steven is a successful poet, novelist, children's writer and broadcaster who has published some twenty-five books. His BBC Radio 4 documentary on the island of St Kilda won him a Sony Award, and he has written and narrated several series of the BBC Radio 3 *Essay* series. His novels include *The Well of the North Wind* (2016), a spiritual tale set on 6th-century Iona, and *2020* (2017), which was longlisted for the Portico Prize. In non-fiction, *Beneath the Ice* (2016) tells the story of the Arctic Sami people. He is also the author of a companion essay collection in the In the Moment series, *Atoms of Delight* (2024).